KHAHARI DISCOVERS

the Brilliance of Janae

Evan J. Roberts

Illustrated by Janine Carrington

Empowered People Press

To my beautiful nieces Kierra and Kourtney.
You are exceptional and brilliant!
Copyright © 2018 by Evan J. Roberts
Editor: Janice Pernell
Cover Design and Illustrations:
Janine Carrington

All Rights Reserved. This book may not be reproduced in whole or in part without the express written consent of the publisher, except by a reviewer who may quote brief passages in a review.
Nor may any part of this book be reproduced, stored in a retrieval system, or transmitted in any form or by any means, electronic, mechanical, photocopying, recording, or other, without the express written permission of the publisher.
Third Series, 2018 Manufactured in the United States of America
Library of Congress Control Number: 2016901009
ISBN: 978-0-9966463-3-8

"Dad, they're here! They're here!" Khahari shouted.
Outside, Aunt Brenda, Uncle Myles, and Janae smiled and waved.

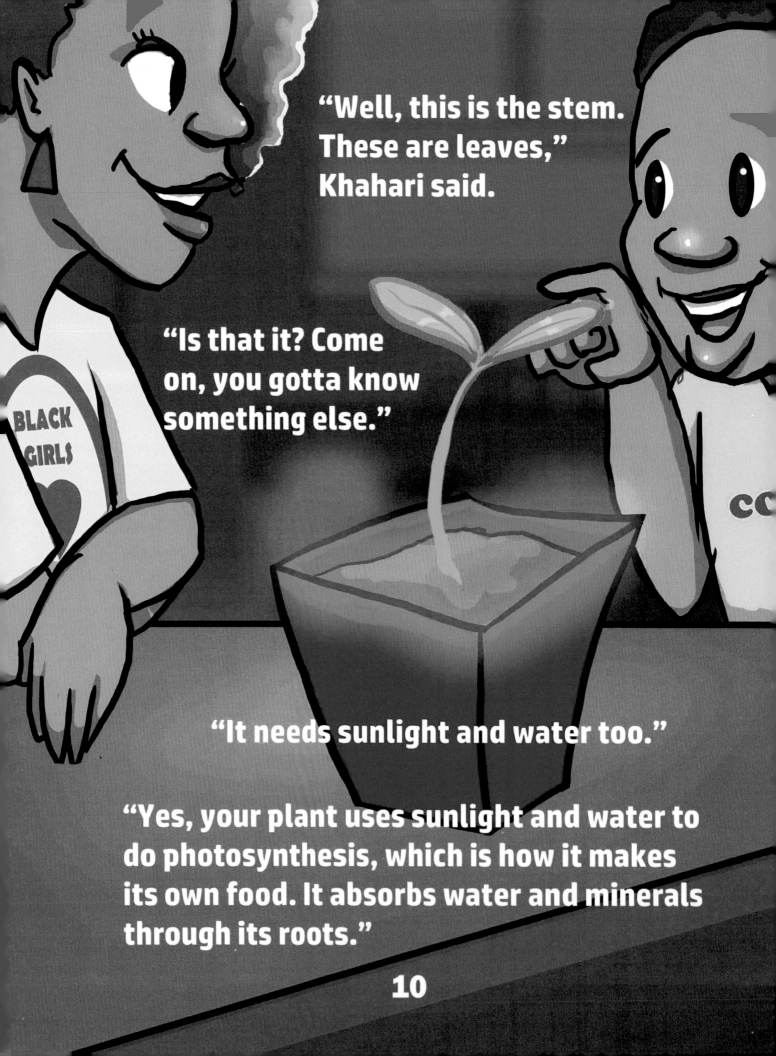

"Wow! I didn't know my plant did all of that. Will you help me take care of it?" Khahari asked.

"Yeah, it's easy. It needs moist soil and plenty of sun," Janae added.

Khahari listened, moving the plant to a nearby window sill

"We'll check and measure it every day."

"Like scientists?" Khahari asked.

"Sure. We can pretend like we're botanists, aka super plant investigators."

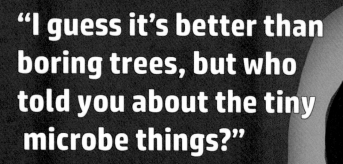

"I guess it's better than boring trees, but who told you about the tiny microbe things?"

"My mom teaches biology. Plus, I want to be a microbiologist like Dr. Carolyn Brooks. She knows all about microbes and growing stuff."

"She taught my STEAM camp last summer at this great school called Tuskegee. Tuskegee has a lot of famous scientists who went there.

We did science, computer stuff, art, and math. We made robots too. Mom wanted me to go. It was a blast!"

"Hey you two! Guess what? Uncle Navey is taking us to the Conservatory."

"Oh, cool!" Janae shouted.

"Is that like the fish aquarium?" Khahari asked.

"More like the plant aquarium, but it's fun."

"Aww, man! How can boring plants be fun?"

"You'll see!

"Guess who was the first black woman to get a Ph.D. in chemistry."

Everybody looked puzzled except Janae.

"It was Dr. Marie Daly of Columbia University."

Then Navey asked, "Guess whose nickname is human computer and who works on spaceships."

"That's Katherine Johnson. She's like a super math genius!" "Did you see the movie about her?"

"Yes we did," Khahari replied.

"It was so good. Black girls are brilliant in science and math too."
"Yes they are! And you're one of the brightest," Navey said.

"All right, the Conservatory's just ahead. Last question. Guess who was the first black woman named Surgeon General."
"Dr. Jocelyn Elders of the University of Arkansas," Janae replied

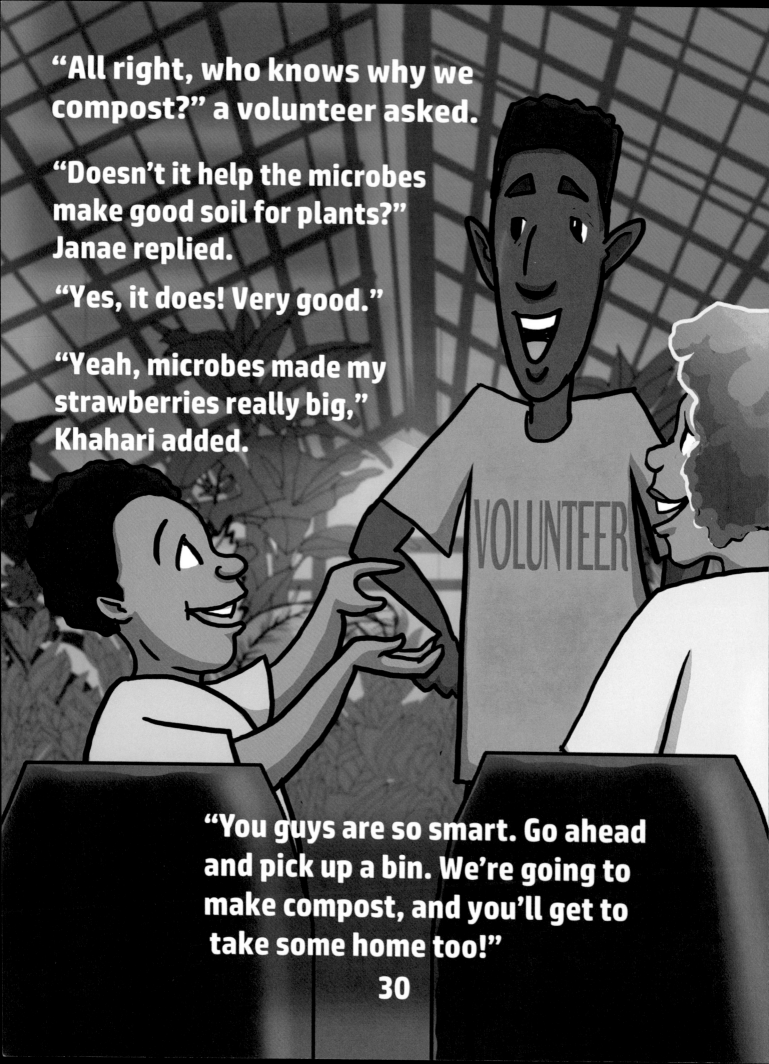

"We need three things to compost: greens, like these grass clippings, browns, like those wood chips, and water," the volunteer said.

Khahari and Janae picked up a pile of greens and browns to put inside their bins.

"Great job! Now that it's set up, the microbes can do their job. In a few weeks, they'll turn this bin into plant food that looks like this," the volunteer said.

"The plant food allows plants to grow and plants help us to live, so make sure you thank a green plant every day."

Made in the USA
Middletown, DE
21 March 2019